For Hugo, Tala,
and Ben

*Spark*an Interest: Allergies A to Z*

For permission requests, write to the author at rachelreyesbrandwood@gmail.com.

NATIONAL LIBRARY OF AUSTRALIA

A catalogue record for this book is available from the National Library of Australia

ISBN: 978-1-7640267-6-5 (Paperback)
ISBN: 978-1-7640267-7-2 (eBook)

Disclaimer: The information presented in this book is intended solely for educational purposes and should not be interpreted as medical advice or an endorsement of any specific medical product, treatment, or procedure. Readers are encouraged to consult a qualified healthcare professional for personalised medical advice or concerns. The author and publisher disclaims any liability arising from the use or reliance on the information contained in this book.

Spark* an Interest

Allergies A to Z

Rachel Reyes-Brandwood

Preface

I wrote this book because I understand firsthand the challenges of living with allergies. As a child, I experienced eczema after eating seafood, but it was not until I became a parent that I truly grasped the complexities of managing allergies within a family. My journey began when my son was diagnosed with life-threatening allergies to milk, egg, peanut, tree nuts, sesame, garlic, and potato at just five months old. Later, my daughter developed respiratory issues, joint pains, and stomach discomfort at age two, which we eventually linked to her consumption of carrots, beans, peas, corn, citrus fruits, apples, etc. Meanwhile, my husband discovered his own sensitivity to wheat and gluten while experimenting with baking bread in response to the ubiquitous "may contain sesame" warnings on store-bought bread in New South Wales. Navigating life with these challenges has deepened my understanding of allergies and their impact on families.

I designed this book to educate families and friends living with or supporting loved ones with allergies. My hope is to make complex concepts accessible while offering practical advice to help others manage allergies safely and confidently. This book is for you, whether you are a parent, caregiver, friend or simply curious about the science behind allergies.

Most of all, I am particularly passionate about fostering allergy awareness, especially for children starting school, where safety and understanding are crucial. By empowering young minds with knowledge about allergies, I hope to create a more inclusive and supportive environment for all students, ensuring that every child can thrive without fear of their allergies holding them back.

Rachel Reyes-Brandwood
1 April 2025

A is for Allergen

An allergen is something that your body thinks is bad for you, even though it's usually harmless to most people. Common allergens include pollen, dust mites, and certain foods like milk, eggs and peanuts. When your body meets an allergen, it can react by causing symptoms like sneezing, itching, or even more serious reactions.

B is for Breathing difficulties

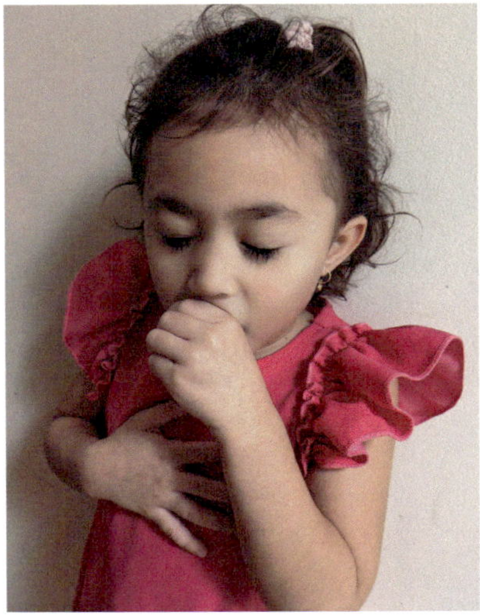

Breathing difficulties can happen when your body reacts to an allergen. This can make the tiny tubes in your lungs tighten, making it harder to breathe. In very serious allergic reactions, called *anaphylaxis*, breathing can become much harder because your throat or tongue might swell too. If you or someone else is having trouble breathing, especially with other symptoms like swelling, dizziness, or a rash, it's important to take quick action! (*See letter* Q.)

C is for Challenge test

A challenge test is a special way doctors check if you're really allergic to a food. They get you to eat tiny amounts of the food in a safe place and watch closely to see how your body reacts, especially when other tests aren't clear. Challenge test can also show if you've outgrown an allergy you used to have, or if you can eat foods like baked egg or milk even if you're allergic to them in other forms.

D is for Desensitisation

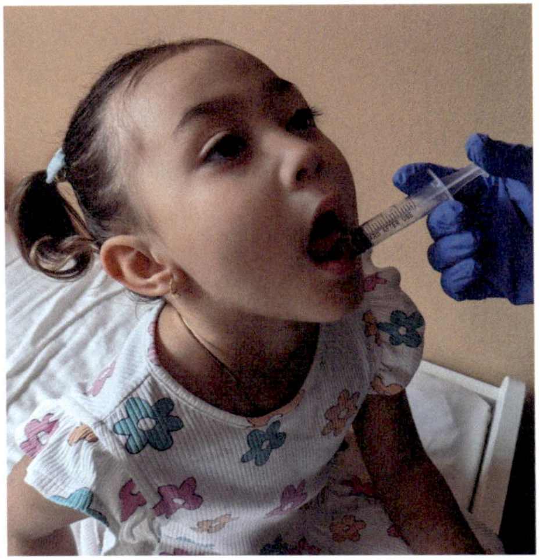

Desensitisation is a treatment that helps your body get used to allergens, like pollen or foods. Over a period of time, you are given super tiny amounts so allergic reactions get weaker. Doctors do it safely as shots (SCIT), drops or tablets under your tongue (SLIT), or by feeding you small bits (OIT), helping your body learn the allergen isn't so scary. This often makes allergy symptoms milder and easier to handle.

E is for Epinephrine (EpiPen)

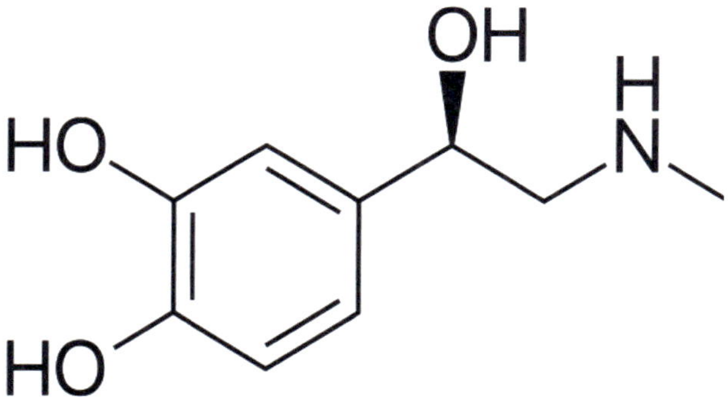

Epinephrine is a special medicine that helps your body fight back when you have a very bad allergic reaction called *anaphylaxis*. It's given as a quick shot on your leg using a device called EpiPen or Anapen, which is a useful pen that can help save your life if you're having a severe allergic reaction. After using EpiPen or Anapen, always call for an ambulance because doctors need to check on you to make sure you're okay, even if you start feeling better.

F is for Food intolerance

Lactose Free

GLUTEN FREE

Food intolerance is when your body has trouble digesting certain foods, like lactose in milk, gluten in wheat, or FODMAP in certain fruits and vegetables, which can cause symptoms like tummy pain, bloating, or diarrhoea/diarrhea. Unlike allergies, food intolerances don't involve the immune system, so they're generally not as serious. They are more about how your body handles certain foods, and you might be able to eat small amounts without problems.

G is for Gluten

Gluten is a special protein found in foods like bread, pasta, and cereal. Some people's bodies don't like gluten and it can make them feel sick or upset. For kids with ceoliac/celiac disease, eating gluten can make their tummy hurt and may even make them feel sad or have trouble paying attention. That's why doctors tell these kids to eat special foods without gluten to help them feel better and healthier.

H is for Histamine

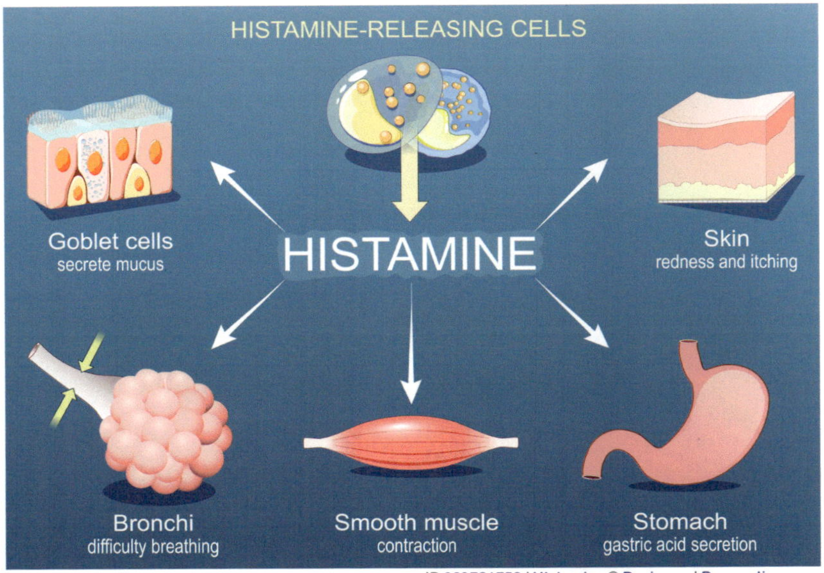

ID 360721753 | Histamine © Designua | Dreamstime.com

Histamine is a special chemical in your body that helps protect you by telling your immune system to fight off infections and also helps with digestion and brain functions, but sometimes it can make you feel uncomfortable. During an allergic reaction, when your body meets an allergen, it releases histamine, which can make you sneeze, have a runny nose, or feel itchy. If too much histamine is released, it can even cause swelling or make it hard to breathe, which is why doctors help treat allergies to keep you safe.

I is for IgE (Immunoglobulin E)

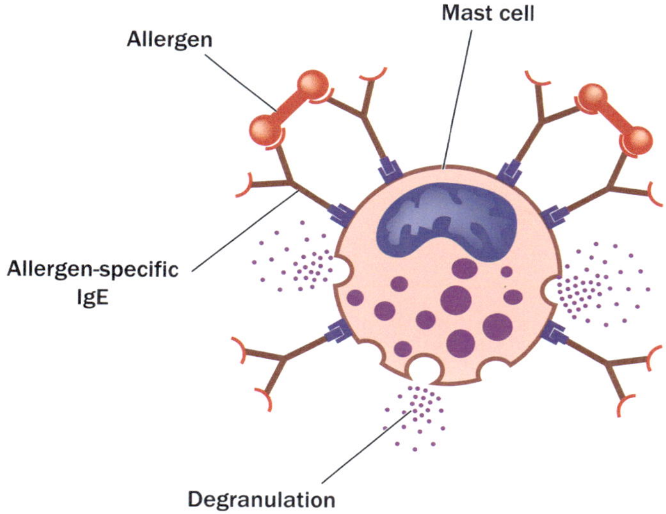

ID 72564939 © Legger | Dreamstime.com

IgE is a special helper in your body that's like a radar looking for harmful things. It attaches to what it thinks is dangerous, like allergens, and tells your body to release special chemicals like histamine, causing symptoms such as sneezing, itching, or swelling. In conditions like asthma and eczema, IgE levels are often higher, and doctors can measure these levels to help diagnose allergies and guide treatment.

J is for Juvenile allergies

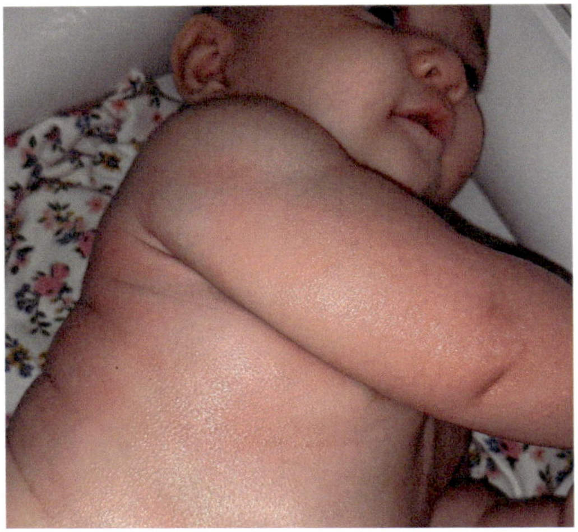

Allergies in children and adolescents happen when the body overreacts to things that are usually harmless like certain foods or things in the air. Some common allergies in kids include rhinitis from pollen or dust (hay fever), itchy skin rashes (eczema), breathing problems that can get worse with allergies (asthma), and food allergies. These allergies can start when you're young, and sometimes they get better as you grow up, but other times they may stick around even when you're older.

K is for Kitchen safety

Kitchen safety for children with food allergies means being careful about what you touch and eat. Always wash your hands, keep safe and unsafe foods separate, use clean utensils and surfaces, read food labels and look for allergen warnings, and always ask a grown-up before trying new foods. Also, make sure everyone around you knows about your allergies so they can help keep you safe.

L is for Lactose

Lactose is a special sugar found in milk and dairy products that some people's bodies have trouble digesting. When someone can't digest lactose well, they might feel tummy pain, get bloated, or have diarrhoea/diarrhea after eating or drinking dairy foods. This is called lactose intolerance which happens when the body doesn't make enough of an enzyme called lactase to break down the lactose sugar.

M is for Mast cells

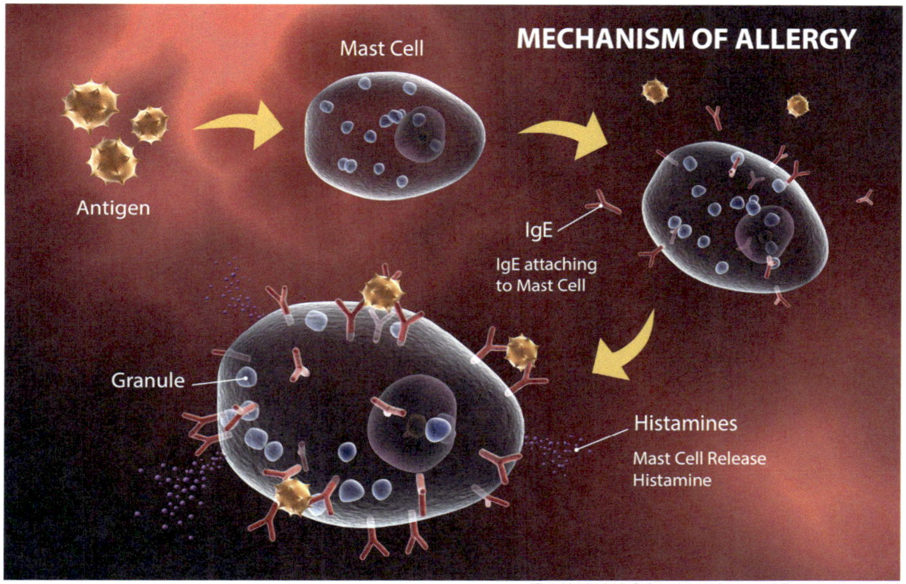

If IgE is the radar that spots allergens and histamine is the special chemical that causes allergy symptoms, mast cells are like the storage units and alarm system in your body. They hold histamine and other chemicals, and when IgE spots an allergen, mast cells release these chemicals to signal your body to react. This can cause symptoms like sneezing, itching, or swelling during an allergic reaction, or even stronger reactions in conditions like MCAS, where mast cells overreact too easily.

N is for Non-IgE-mediated allergy

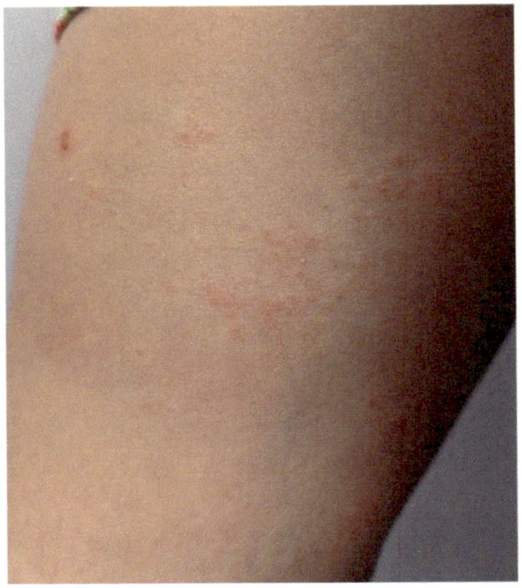

Non-IgE-mediated allergy (sensitivity) are a special kind of allergy that can make your tummy upset or your skin itchy (eczema), but they work differently from other allergies. Instead of causing quick reactions like sneezing or swelling, these allergies might make you feel sick hours or even days after eating certain foods. Doctors can't use the usual allergy tests to find them, so they might ask you to stop eating some foods for a while and then try them again to see if they're causing your symptoms.

O is for Oral allergy syndrome (OAS)

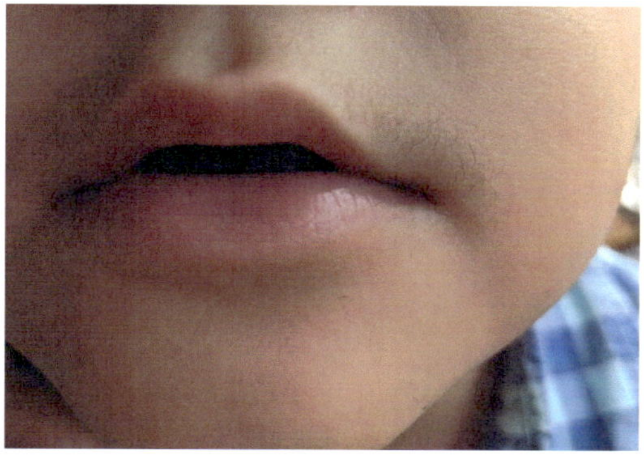

Oral allergy syndrome is when your mouth gets itchy or tingly after eating certain fresh fruits or veggies, like apples or carrots. It happens because your body thinks these foods are like the pollen you're allergic to, even though they're not really the same thing. Usually, it's not too serious and only affects your mouth, but in rare instances, it could get worse. Therefore, it's important to tell a grown-up if it happens to you so they can help you figure out which foods to be careful with.

P is for Pollen count

Pollen is a fine, yellow powder that plants make to help them reproduce, and it can get into our noses and eyes when it's in the air. A pollen count measures how much pollen is in the air on a given day, which is important for people with hay fever or seasonal allergies because high counts can make symptoms worse. By checking the pollen count and wearing a mask when outside, you can help manage your symptoms and feel better on days when pollen levels are high.

Q is for Quick action (importance of)

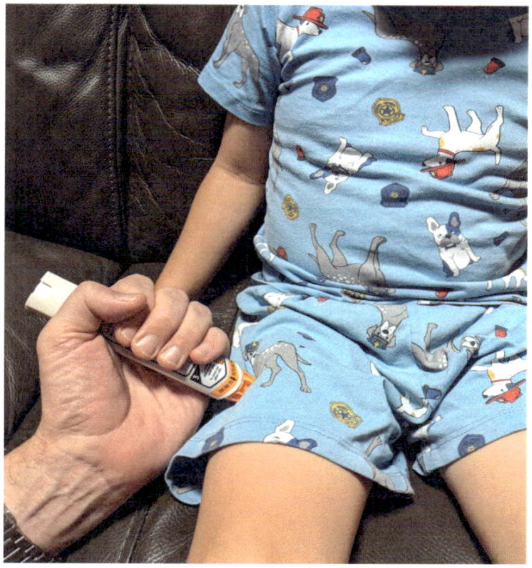

Severe allergic reactions can be life-threatening, so quick action is super important if you're having one. For mild symptoms like itching or rashes, you can use antihistamine or corticosteroid creams. But if you're having trouble breathing or feeling very sick, you need to use EpiPen or Anapen right away. It works fast to help you breathe better and feel better overall, but you still need to go to the hospital after using it to make sure you're okay.

R is for Rhinitis (allergic rhinitis)

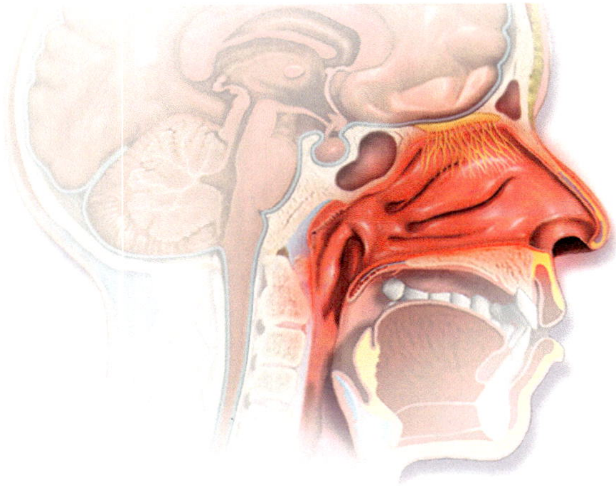

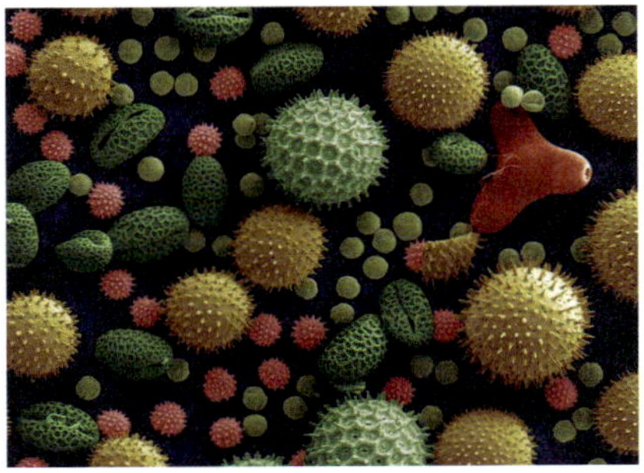

Rhinitis is when your nose gets stuffy, runny and itchy because of allergies. It happens when your nose meets things it doesn't like, such as pollen (then you have hay fever), dust mites, or pet dander, and decides to overreact. As a result, you might sneeze, feel like your nose is blocked, or have a runny nose, and sometimes your eyes might get itchy and watery too.

S is for Skin prick test

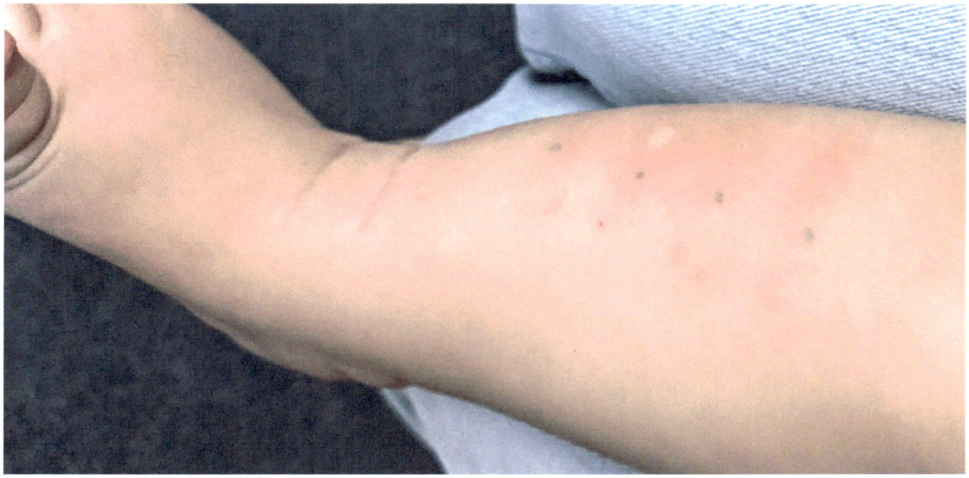

A skin prick test is a way for doctors to check if you're allergic to something by putting tiny drops of different substances on your skin and gently scratching it with a small needle. If you're allergic, a little red, itchy bump might appear where the drop was, kind of like a mosquito bite. The size of this bump can give doctors an idea about how sensitive you are to that allergen: a bigger bump usually means you're more sensitive, but it doesn't always mean your allergy symptoms will be worse.

T is for Tree nuts

Tree nuts are a kind of food that grows on trees like almonds, Brazil nuts, cashews, hazelnuts, macadamia nuts, pecans, pine nuts, pistachios and walnuts. Some people have allergies to tree nuts, which means their body reacts badly when they eat them, causing symptoms like hives, tummy pain, or trouble breathing. If you have a tree nut allergy, it's important to be careful about what you eat and always ask a grown-up before trying new foods because they may contain nuts.

U is for Urticaria (hives)

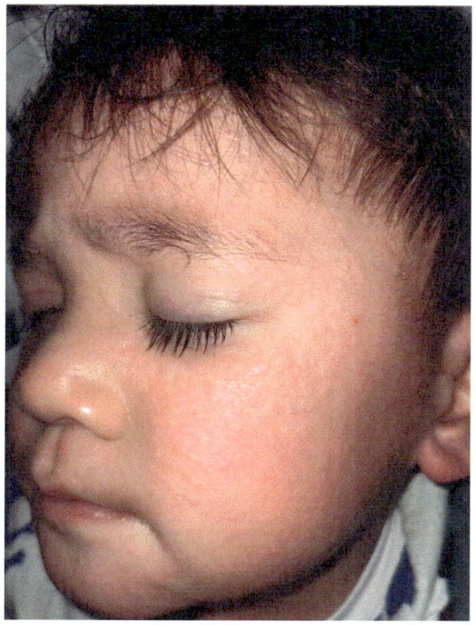

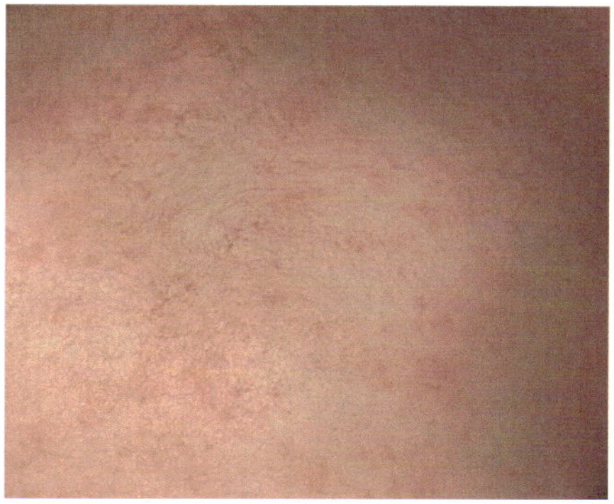

Urticaria, also known as hives, are itchy, red bumps that can appear on your skin when your body reacts to something it doesn't like, such as certain food or insects. They can also be caused by things like being too cold or too hot. Sometimes, urticaria can be a sign of a more serious reaction, so it's important to tell an adult if you have trouble breathing, feel very sick, or if the hives cover a large part of your body.

V is for Venom

Venom is a special liquid that some insects, like bees and wasps, and animals put into your body when they sting or bite you. For most kids, it just hurts a little, but for some, their body thinks the venom is really dangerous and reacts too strongly. This can cause severe allergic reaction, making them very sick with symptoms like swelling or trouble breathing.

W is for Wheezing

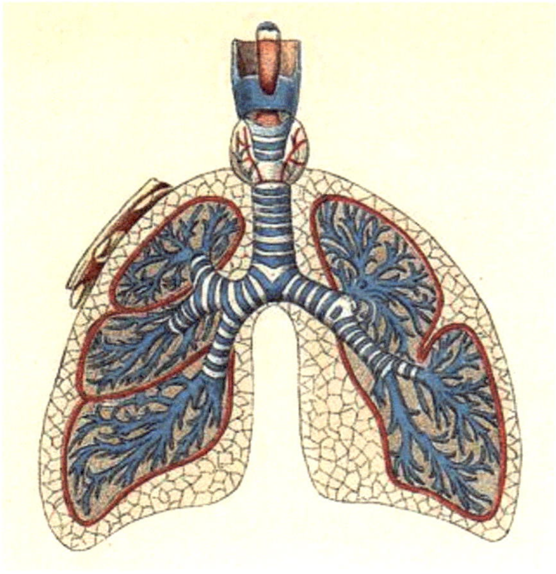

Wheezing is a whistling sound you might hear when you breathe, especially when breathing out. It happens when the tiny tubes in your lungs that help you breathe become narrow, making it harder for air to move through. Sometimes, allergies can cause these tubes to narrow, leading to the whistling sound, kind of like when you blow through a small straw.

X is for Xerostomia

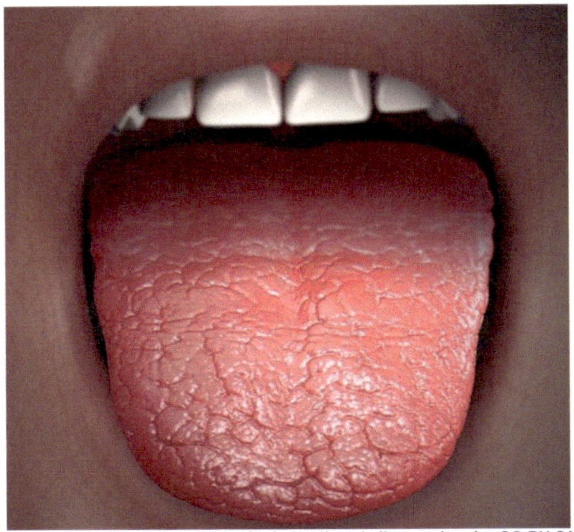

“Xerostomia – Dry Mouth” by Scientific Animations is licensed under CC-BY-SA-4.0.
[http://www.scientificanimations.com/wiki-images/]

Xerostomia, or dry mouth, is when your mouth feels very dry because it doesn’t have enough spit (saliva). Sometimes, allergies can make your nose stuffy, causing you to breathe through your mouth, which can dry it out. Also, some medicines used to treat allergies can make your mouth feel dry as a side effect.

Y is for Yeast

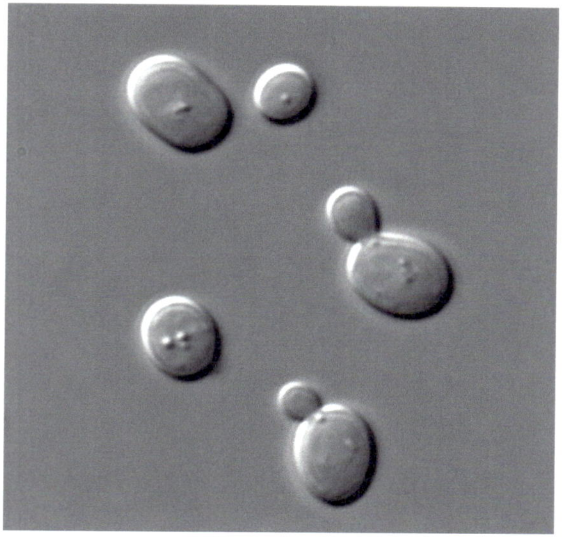

Yeast is a tiny fungus that lives naturally in your body and is also found in foods like bread and cheese. Sometimes, people can be allergic to yeast, which means their body reacts to it as if it's harmful. This can cause tummy problems like bloating or diarrhoea/diarrhea, itchy skin, or even sneezing and breathing issues.

Z is for Zonulin

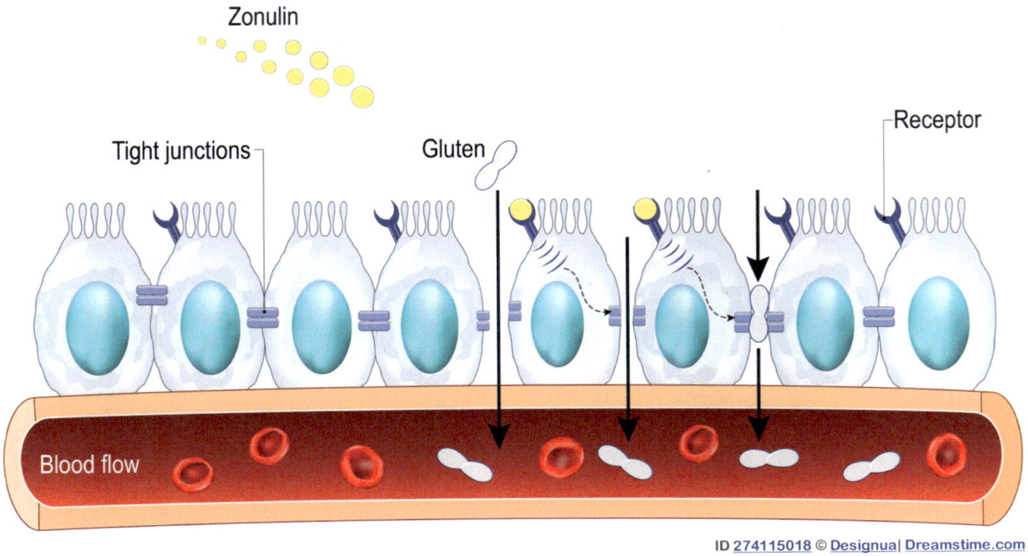

Zonulin is a special protein in your gut that's like a doorkeeper, controlling tiny doors between cells in your intestines to let good things in and keep bad things out. Sometimes, zonulin can get confused and open the doors too much, which might make you feel sick or cause allergies. In people with ceoliac/celiac disease, eating gluten makes zonulin open these intestinal doors too wide, letting gluten in and causing problems, which is why they need to avoid gluten to stay healthy.

Description of Images

A: Photographs of common allergens such as (clockwise) tree pollen, milk, peanuts, egg, and dust mite

B: Photograph of a five year-old child exhibiting difficulty in breathing

C: Photograph of food samples prepared for a tree nuts challenge test

D: Photograph of a re-enactment of a child getting sublingual immunotherapy (SLIT)

E: Diagram of the molecular structure of epinephrine (left) and photograph of an epinephrine auto-injector, EpiPen (right)

F: Gluten free and lactose free labels, and photographs of foods high in FODMAP (Fermentable Oligosaccharides, Disaccharides, Monosaccharides, and Polyols) such as beans, apple, and broccoli. These are short-chain carbohydrates that are poorly absorbed in the small intestine and ferment in the colon, producing gas and potentially causing digestive discomfort in some individuals.

G: Photograph of foods with gluten such as pasta, bread, and cereals

H: Diagram illustrating histamine's central role in mediating allergy symptoms and immune response. Histamine triggers various physiological changes including: increased gastric acid secretion of the stomach, contraction of smooth muscle which can affect organ function, difficulty breathing due to airway constriction of the bronchi, enhanced mucus secretion of goblet cells, and redness and itching of the skin caused by increased blood flow and inflammation.

I: Diagram depicting allergen-specific IgE binding to allergens on the surface of mast cell, leading to the activation and degranulation of mast cells. This process results in the release of vasoactive amines including histamine, which play a significant role in allergic reaction and inflammation.

J: Photograph of a four month-old baby experiencing an allergic reaction after skin contact with milk

K: Photographs of handwashing (left) and food labelling as safe and unsafe (right)

L: Photograph of dairy foods such as cheeses and milk

M: Diagram illustrating the mechanism of an allergic reaction. Antigens (allergens) bind to IgE antibodies attached to mast cells, triggering mast cells to release granules containing histamine and other chemicals.

These substances cause symptoms such as inflammation, itching, and airway constriction, which are characteristic of allergic responses.

N: Photograph of eczema on a child’s leg

O: Photograph of a three year-old child’s mouth

P: Photograph of pollen of a yellow flower

Q: Photographs of an allergy medication case (left) and an epinephrine auto-injector training device on a three year-old child’s thigh (right), demonstrating the quick action needed during severe allergic reactions

R: Illustration of rhinitis (left), highlighting inflammation in the nasal passages, redness and swelling of the mucous membranes, increased mucus production, and irritation of nerve endings, which can lead to symptoms such as congestion, sneezing, and nasal discharge; and illustration of various pollens magnified 500 times (right)

S: Photograph of skin prick test on a leg of a one year-old child

T: Photograph of various tree nuts

U: Photographs of urticaria on face (left) and back (right) of a two year-old child from a change in temperature

V: Photographs of a bee on petal (left) and a wasp on human hand (right)

W: Illustration of lungs (left) and photograph of a whistle (right), symbolising the whistling sound coming from the narrowed airways in the lungs when wheezing

X: Illustration of xerostomia (dry mouth), showing a cracked and dehydrated tongue, a common symptom caused by reduced saliva production

Y: Photographs of dry yeast on a teaspoon (left) and yeast under a microscope (right)

Z: Illustration of zonulin’s role in intestinal permeability, showing gluten triggering zonulin release, which disrupts tight junctions in the intestinal lining, allowing particles to pass into the bloodstream

Made in United States
Troutdale, OR
11/03/2026

56844715R00021